# PROBLEM SOLVING AND REASONING STUDENT'S BOOK 1

Peter Clarke

William Collins' dream of knowledge for all began with the publication of his first book in 1819. A self-educated mill worker, he not only enriched millions of lives, but also founded a flourishing publishing house. Today, staying true to this spirit, Collins books are packed with inspiration, innovation and practical expertise. They place you at the centre of a world of possibility and give you exactly what you need to explore it.

Collins. Freedom to teach.

Published by Collins
An imprint of HarperCollins*Publishers*
The News Building
1 London Bridge Street
London
SE1 9GF

HarperCollins*Publishers*
1st Floor, Watermarque Building, Ringsend Road
Dublin 4, Ireland

Browse the complete Collins catalogue at
**www.collins.co.uk**

ISBN 978-0-00-827177-0

The author wishes to thank Brian Molyneaux for his valuable contribution to this publication.

British Library Cataloguing in Publication Data
A Catalogue record for this publication is available from the British Library

Author: Peter Clarke
Publishing manager: Lizzie Catford
Editors: Amy Wright, Mike Appleton
Copyeditor: Catherine Dakin
Proofreader: Tanya Solomons
Answer checker: Steven Matchett
Cover designer: Ink Tank, Ken Vail Graphic Design Ltd
Cover image: Rob Hainer/Shutterstock
Internal designer: Ken Vail Graphic Design Ltd
Typesetter: Ken Vail Graphic Design Ltd
Illustrator: Eva Sassin
Production controller: Sarah Burke
Printed and bound in the UK using 100% Renewable Electricity at CPI Group (UK) Ltd

# Contents

## Solving mathematical problems

Please note that Strand 5 of the Cambridge Primary Mathematics Curriculum Framework: Problem solving is incorporated into each challenge.

# Reasoning mathematically

Please note that Strand 5 of the Cambridge Primary Mathematics Curriculum Framework: Problem solving is incorporated into each challenge.

# Using and applying mathematics in real-world contexts

| | Cambridge Primary Mathematics Curriculum Framework | | | | | | | |
| --- | --- | --- | --- | --- | --- | --- | --- | --- |
| | Number – Numbers and the number system | Number – Calculation: Addition and subtraction, including Mental strategies | Number – Calculation: Multiplication and division, including Mental strategies | Geometry – Shapes and geometric reasoning | Geometry – Position and movement | Measure – Money ($) Length (L), Mass (M), Capacity (C), Time (T) | Handling data | |
| Finding numbers | • | | | | | | • | 56 |
| List of numbers | • | | | | | | • | 57 |
| Card patterns | • | | | | | | • | 58 |
| Types of shoes | • | | | | | | • | 59 |
| Jelly beans | • | | | | | | • | 60 |
| Favourite bear | • | | | | | | • | 61 |
| Flags | • | | | | | | • | 62 |
| Letters in a name | • | • | | | | | • | 63 |
| Greg's garage | • | | • | | | | • | 64 |
| Takeaway food | • | | • | | | | • | 65 |
| Recycling | • | • | • | | | | • | 66 |
| Things with wheels | • | | | • | | | • | 67 |
| Pizza box | | | | • | | | | 68 |
| Things that move | | | | • | • | | • | 69 |
| Turning letters | | | | | • | | | 70 |
| Treasure | | | | | • | | | 71 |
| Name's worth | • | • | | | | • ($) | • | 72 |
| Beads | • | | | | | • (L) | | 73 |
| How tall? | • | | | | | • (L) | | 74 |
| Newspaper | • | | | | | • (L) | | 75 |
| Sheet of paper | • | | | | | • (L/M) | | 76 |
| Yoghurt pots | | | | | | • (M/C) | • | 77 |
| Jobs rota | • | | | | | • T | • | 78 |
| Cover it | • | | | | | • (Area) | | 79 |
| When you've finished, … | | | | | | | | 80 |

Please note that Strand 5 of the Cambridge Primary Mathematics Curriculum Framework: Problem solving is incorporated into each challenge.

# How to use this book

## Aims

This book aims to provide teachers with a resource that enables learners to:

- develop mathematical problem solving and thinking skills
- reason and communicate mathematically
- use and apply mathematics to solve problems.

## The three different types of mathematical problem solving challenge

This book consists of three different types of mathematical problem solving challenge:

Problem solving and mathematical reasoning

**Solving mathematical problems**

This involves learners investigating, exploring and applying their mathematical knowledge and skills to solve problems 'within' mathematics itself.

**Reasoning mathematically**

This involves learners using logical thinking to solve problems, focusing on making conjectures and generalisations, and explaining and justifying conclusions using appropriate language.

**Using and applying mathematics in real-world contexts**

This involves learners engaging in challenges that require them to use and apply their mathematical knowledge and skills in open-ended, real-world contexts.

This book is intended as a 'dip-in' resource, where teachers choose which of the three different types of challenge they wish learners to undertake. A challenge may form the basis of part of or an entire mathematics lesson. The challenges can also be used in a similar way to the weekly bank of 'Additional practice activities' found at the end of each unit in the Collins International Primary Maths Teacher's Guide. It is recommended that learners have equal experience of all three types of challenge during the course of a term.

The 'Solving mathematical problems' and 'Reasoning mathematically' challenges are organised under the different strands of the Cambridge Primary Mathematics Curriculum Framework. This is to make it easier for teachers to choose a challenge that corresponds to the topic they are currently teaching, thereby providing an opportunity for learners to practise their pure mathematical knowledge and skills in a problem solving context. These challenges are designed to be completed during the course of a lesson.

The 'Using and applying mathematics in real-world contexts' challenges have not been organised by strand. The very nature of this type of challenge means that learners are drawing on their mathematical knowledge and skills from several strands in order to investigate challenges arising from the real world. In many cases these challenges will require learners to work on them for an extended period, such as over the course of several lessons, a week or during a particular unit of work. An indication of which strands each of these challenges covers can be found on page 5.

Please note that Strand 5 of the Cambridge Primary Mathematics Curriculum Framework: Problem solving is incorporated into each challenge.

## Briefing

As with other similar teaching and learning resources, learners will engage more fully with each challenge if the teacher introduces and discusses the challenge with the learners. This includes reading through the challenge with the learners, checking prerequisites for learning, ensuring understanding and clarifying any misconceptions.

## Working collaboratively

The challenges can be undertaken by individuals, pairs or groups of learners, however they will be enhanced greatly if learners are able to work together in pairs or groups. By working collaboratively, learners are more likely to develop their problem solving, communicating and reasoning skills.

## You will need

All of the challenges require learners to use pencil and paper. Giving learners a large sheet of paper, such as A3 or A2, allows them to feel free to work out the results and record their thinking in ways that are appropriate to them. It also enables learners to work together better in pairs or as a group, and provides them with an excellent prompt to use when sharing and discussing their work with others.

An important problem solving skill is to be able to identify not only the mathematics, but also what resources to use. For this reason many of the challenges do not name the specific resources that are needed.

## Characters

The characters on the right are the teacher and the four children who appear in some of the challenges in this book.

Mr Lewis     Holly     Jake     Ruby     Meru

## Think about …

All challenges include prompting questions that provide both a springboard and a means of assisting learners in accessing and working through the challenge.

## What if?

The challenges also include an extension or variation that allows learners to think more deeply about the challenge and to further develop their thinking skills.

## When you've finished, …

At the bottom of each challenge, learners are instructed to turn to page 80 and to find a partner or another pair or group. This page offers a structure and set of questions intended to provide learners with an opportunity to share their results and discuss their methods, strategies and mathematical reasoning.

When you've finished, turn to page 80.

## Solutions

Where appropriate, the solutions to the challenges in this book can be found at *Collins International Primary Maths* on Collins Connect and on our website: collins.co.uk

# 12 in the middle

## Challenge

**12**

Write a number pattern that has 12 in the 3rd position.

How many different patterns can you make with 12 in the middle?

## Think about ...

Think about patterns that count on and patterns that count back.

What pattern is your sequence going to follow? What is the rule for your pattern?

## What if?

What if the number 30 is in the middle of your number pattern?
How many different patterns can you make?

**30**

When you've finished, turn to page 80.

# Showing numbers

## Challenge

You have 3 tens rods and 5 ones blocks.

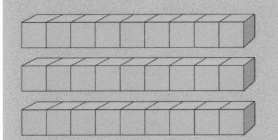

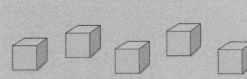

Using some of these rods and blocks, you can make the number 23.

What other numbers can you make using some or all of the rods and blocks?

| Tens | Ones |
|------|------|

**You will need:**
- Base 10 rods and blocks – tens and ones
- abacus

## Think about ...

What patterns do you notice as you make your numbers? Can this help you to make other numbers?

Think about using only tens rods or only ones blocks.

## What if?

This abacus shows the number 23 using 5 beads.

Write all the numbers you can make on an abacus using 5 beads.

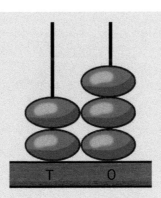

T    O

When you've finished, turn to page 80.

# Ladybird counting

## Challenge

This ladybird must have the same number of spots on both sides of its back.

Which numbers of counters work? Why?

Which numbers of counters don't work? Why not?

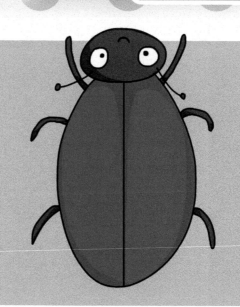

**You will need:**
• small counters

## Think about ...

What pattern do you notice for the numbers of counters that work? What about the numbers that don't work?

Without using counters, can you predict which numbers will work and which numbers will not?

## What if?

What if there are two ladybirds, and each ladybird must have the same number of spots on both sides of its back?

Which numbers work? Why? Which numbers don't work? Why not?

When you've finished, turn to page 80.

# 2-digit numbers

## Challenge

How many different 2-digit numbers can you make using these cards?

Each card can only be used once in each number.

**You will need:**
- two sets of 0–6 digit cards (optional)

4    3    1    6    2

## Think about ...

What is the largest number you can make? What is the smallest number?

Write your numbers in order, from smallest to largest. Have you missed out any numbers? How do you know? What are the numbers you have missed?

## What if?

What numbers can you make between 14 and 34 using these cards?
This time, you can use the same card twice to make a number.

2    5    3    1    0

When you've finished, turn to page 80.

# Half a shape

## Challenge

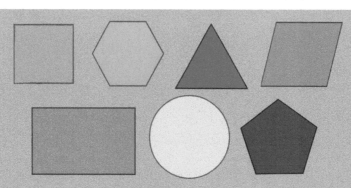

**You will need:**
- set of 2-D shapes
- ruler

Take a 2-D shape and trace around it.

How many different ways can you divide the shape in half?

Use different tracings of the shape to show each way of dividing it in half.

Repeat for several shapes.

## Think about ...

Make sure that your drawings are as accurate as possible. Why is this important?

Which shapes are easy to divide in half? Why is this? Which shapes are not so easy? Why?

For the 'What if?', make sure that the sides of both **half shapes** touch completely.

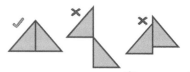

How many different ways can you divide each of these shapes into four equal parts?

## What if?

Each of these shapes is half of another shape.

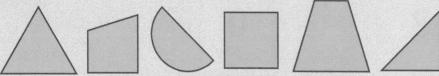

Investigate what each **whole shape** might look like.

When you've finished, turn to page 80.

12

# Half the counters

## Challenge

Take a handful of counters and count them.

Can you halve the counters by arranging them into two equal groups?

Take other handfuls of counters and repeat.

Which numbers can be halved?

Which numbers cannot be halved?

**You will need:**
• counters

## Think about ...

What do all the numbers that can be halved have in common?

What do all the numbers that cannot be halved have in common?

## What if?

Look at these shapes made from counters. What do you notice about the numbers of counters in the shapes? How many counters will the next shape have? How do you know?

What do you notice about the numbers of counters in these shapes? How many counters will the next shape have? How do you know?

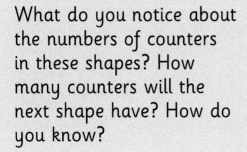

When you've finished, turn to page 80.

13

# Scoring 12

## Challenge

Which pins do you need to knock down to make a total of 12?

How many different ways can you do it?

## Think about ...

You can knock down two or more pins at a time.

How are you going to record the different ways of scoring 12?

## What if?

What is the largest score you can get by knocking down two pins?

What is the smallest score you can get?

When you've finished, turn to page 80.

# Differences

## Challenge

This domino has a difference of 3.

How many other dominoes can you find with a difference of 3?

Which dominoes have a difference of 2?

Which have a difference of 4?

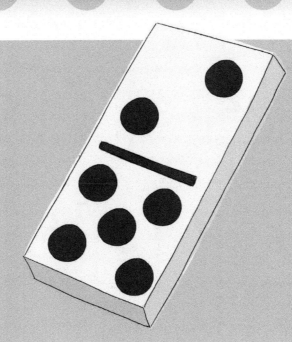

**You will need:**
• set of dominoes
• two 1–6 dice
• two 0–9 dice

## Think about ...

How will you know when you have found all the dominoes with a difference of 3? What about those with a difference of 2 or 4?

How might you write a domino with a difference of 3 as a subtraction number sentence?

## What if?

This pair of 1–6 dice has a difference of 2.

Which other pairs of 1–6 dice have a difference of 2?

What if you use a pair of 0–9 dice?

When you've finished, turn to page 80.

# 3-card number sentence

## Challenge

**You will need:**
- 1–9 digit cards
- +, – and = cards (optional)

Three of these 1–9 digit cards make an addition number sentence:

$5 + 3 = 8$

Using only the 1–9 digit cards, how many different addition number sentences can you make?

## Think about …

> How many different number sentences can you make with a total of 8? What about a total of 1? 2? 3? …?

> What patterns can you spot to help you write as many different number sentences as possible?

## What if?

How many different subtraction number sentences can you make using the 1–9 digit cards?

When you've finished, turn to page 80.

# Making 10

## Challenge

How many different ways can you make 10 using this model?

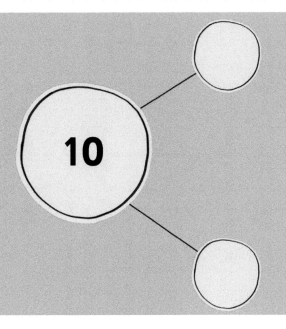

## Think about …

Look for patterns when choosing two or three numbers.

How are you going to record the different ways you can make 10?

Look carefully to make sure that you don't repeat a model.

## What if?

How many different ways can you make 10 using this model?

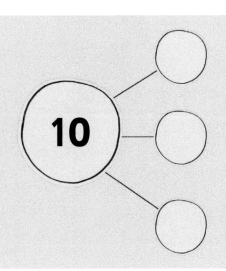

When you've finished, turn to page 80.

# Doubling

## Challenge

1  2  3  4  5  6  7  8  9

Double each of the numbers 1 to 5.

What do you notice about your answers?

Now divide each answer by 2.

What do you notice about these answers?

## Think about ...

How might you record your answers? Try drawing a number line with your answers under the line.

What are you doing when you double a number? What about when you divide a number by 2?

## What if?

What if you double the numbers from 6 to 10?
Do you notice the same things?

When you've finished, turn to page 80.

# Not quite doubles

## Challenge

Choose a number card.

Double your number and add 1.

Double your number and take away 1.

Repeat choosing other number cards

What do you notice about all of your answers?

## Think about ...

Think about how you're going to record your results.

How does doubling a number and adding or subtracting 1 help you work out the answer to 5 + 6 or 3 + 4?

## What if?

What happens if you choose larger numbers?

When you've finished, turn to page 80.

# 24 cubes

## Challenge

Find out how many different ways you can share 24 cubes into equal groups.

How many groups are there?

How many cubes are there in each group?

**You will need:**
- about 40 interlocking cubes (or similar)

## Think about ...

How are you going to record the different ways you found?

Can you work out different ways without having to use the cubes?

## What if?

What if you use 12 cubes? Or 36 cubes?

What other numbers can you share into equal groups?
How many groups are there?
How many cubes are there in each group?

When you've finished, turn to page 80.

# Describing rectangles

## Challenge

Take a sheet of squared paper and cut it into different-sized rectangles. Each rectangle must be made up of at least 6 squares.

**You will need:**
• 2 cm squared paper
• scissors
• counters

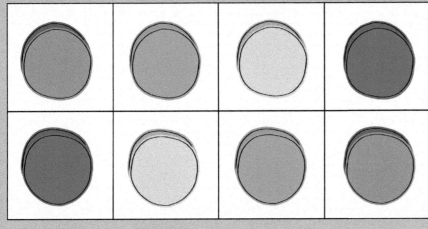

Take one of your rectangles and place a counter in each square.

Describe the rectangle.

Do the same for each of your rectangles.

2 lots of 4 is 8.
4 lots of 2 is 8.

## Think about ...

Think about how you are going to record your descriptions.

Can you describe your rectangles using the language of division?

## What if?

Cut out a large square made from the sheet of squared paper. Describe the square.

When you've finished, turn to page 80.

# Triangular shapes

## Challenge

Fold and cut the square into four triangles.

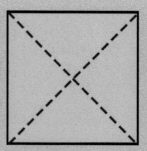

Fit two of the triangles together to make a new shape.

Make sure that the sides of both triangles touch completely.

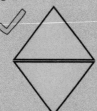 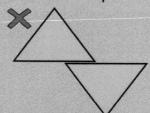

Draw the new shape you have made.

How many different shapes can you make using three triangles?

## Think about ...

Make sure that you don't make the same shape more than once. The same shape can look different when turned around or reflected in a mirror.

## What if?

Investigate how many different shapes you can make by joining four triangles together.

How many different shapes can you make with a square sheet of paper cut into four smaller squares?

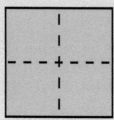

When you've finished, turn to page 80.

# Making cuboids from cubes

## Challenge

9 interlocking cubes have been used to make this set of 4 cuboids.

Using 9 cubes, what other sets of cuboids can you make?

Draw the different cuboids you have made.

Can you use all 9 cubes to make a different set of cuboids?

**You will need:**
- 12 interlocking cubes

## Think about ...

Think about how you are going to show in your drawing how many interlocking cubes you used to make your cuboid.

Think about making 'layers' of cubes to make larger cuboids.

## What if?

What if you use 12 interlocking cubes? How many different cuboids can you make using all 12 cubes?

Can you use all 12 cubes to make different sets of cuboids?

Why can't you make a sphere, cylinder or cone out of the interlocking cubes?

When you've finished, turn to page 80.

# Symmetrical patterns

## Challenge

Place 6 counters on the spots to make a symmetrical pattern about the line of symmetry.

**You will need:**
- paper
- 6 counters all the same colour
- pencil the same colour as the counters

Record your pattern.

Investigate making different patterns using 6 counters.

## Think about ...

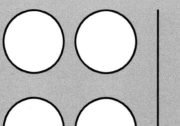

Think carefully about how you're going to record your different patterns. You need to include the spots that you don't place a counter on.

Make sure that you make different patterns and that you don't repeat a pattern.

## What if?

You can't make a symmetrical pattern using 5 counters.

Is Jake right? Explain your answer.

When you've finished, turn to page 80.

# Route

## Challenge

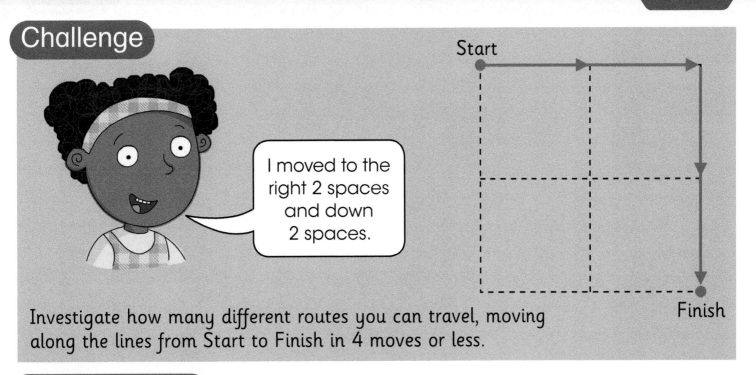

I moved to the right 2 spaces and down 2 spaces.

Investigate how many different routes you can travel, moving along the lines from Start to Finish in 4 moves or less.

## Think about ...

Think about how you will record the different routes you can travel.

Use words such as **right** and **down**, and for the 'What if?' the words **left** and **up** too.

## What if?

Investigate different routes from Start to Finish on this grid.

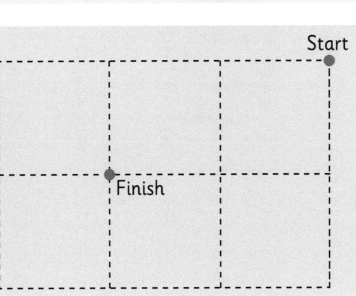

When you've finished, turn to page 80.

25

# Making totals

## Challenge

**You will need:**
• 1c, 5c and 10c coins (optional)

I have two coins.

If Holly has any two of the coins above, what are all the different totals she can make?

## Think about ...

You can use two of the same coin, for example, the two 1c coins.

Think about how you're going to record your different totals.

## What if?

What if Holly has three coins?

What if she has four coins?

When you've finished, turn to page 80.

# Make a metre

## Challenge

Take one set of objects.

Estimate how many will measure 1 metre when you place them in a row.

Now lay that number of objects in a row.

Use the metre stick to see if you're right.

Repeat for the other sets of objects.

**You will need:**
- metre stick
- sets of uniform, non-standard measuring objects, e.g. pencils, paper clips, interlocking cubes

## Think about ...

Make sure that you line up the objects in a straight row, touching slightly.

Try to be as accurate as you can when estimating and measuring.

## What if?

Which set of objects was best for making a metre length? Why was this?

Which objects would be best for measuring something a metre high? Why is this?

When you've finished, turn to page 80.

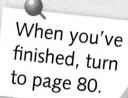

# Comparing objects

## Challenge

Write as many different statements as you can to compare the weights of the four toys.

## Think about …

Use words such as **lighter**, **lightest**, **heavier** and **heaviest**.

Think about comparing pairs of objects. What about comparing more than two objects?

## What if?

How many building blocks weigh the same as the aeroplane?

How many building blocks weigh the same as the train?

The car weighs the same as 3 building blocks, and the fire engine weighs 4 building blocks.

When you've finished, turn to page 80.

# Jugs, bottles and glasses

## Challenge

The bottle holds 4 cups of water, and the jug holds 12 cups of water.

How many different ways can you fill the jug?

## Think about ...

Cups and bottles need to be full when filling the jug.

Think about only using the cup to fill the jug. What if you only use the bottle?

What if you use the cup and the bottle to fill the jug?

## What if?

What if you have to fill two jugs?

When you've finished, turn to page 80.

# Meru's diary

## Challenge

The day after our class trip to the zoo, it's our school's award night.

On Sunday 12th July we're going to visit Nan and Pop.

| July | | | | | | |
|------|--------|-----------|----------|--------|----------|--------|
| **Monday** | **Tuesday** | **Wednesday** | **Thursday** | **Friday** | **Saturday** | **Sunday** |
| | | 1 | 2<br>4:00 Swimming lesson | 3 | 4<br>7:30 Aunty Jane's for supper | 5 |
| 6 | 7<br>Dad away for work  . | 8 | 9 | 10<br>Dad back home | 11<br>10:00 Football match (Home) | 12<br>Visit Nan and Pop |
| 13 | 14<br>Jake's birthday | 15 | 16<br>4:00 Swimming lesson | 17 | 18<br>School fete | 19 |
| 20<br>Class trip to the zoo | 21<br>6:30 School awards night | 22<br>5:30 Dentist ☹ | 23 | 24<br>School holidays start ☺!! | 25<br>10:00 Football match (Away) | 26<br>Family picnic |
| 27 | 28<br>Lucy and Alex over to play | 29 | 30<br>4:00 Swimming lesson | 31 | | |

Write as many different statements as you can that describe the events on the calendar.

## Think about ...

Use words such as: **before, after, next, morning, afternoon** and **evening**.

Write the dates for some of the events.

## What if?

Imagine that today is 23rd July. Write some statements using terms such as: **yesterday, tomorrow, this week, this weekend, last week, last weekend, next week, next weekend**.

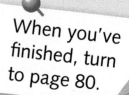

When you've finished, turn to page 80.

# Card sort

Solving mathematical problems

## Challenge

How many different ways can you sort a pack of playing cards?

**You will need:**
- pack of playing cards

## Think about ...

There are 52 cards in a pack of playing cards. If this is too many cards to sort then choose fewer cards, say 20 or 30.

If you're going to sort all 52 cards, make sure that you work in a pair or group.

Think carefully about how you're going to record the different ways you sort the cards.

## What if?

Sort the cards into a simple Venn diagram. Can you create more than one simple Venn diagram?

What about sorting the cards using simple Carroll diagrams?

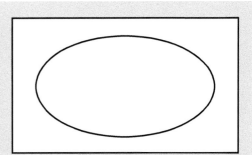

When you've finished, turn to page 80.

# Which is larger?

## Challenge

A number with 8 ones is larger than a number with 5 ones.

Is Holly's statement always true, sometimes true or never true?

Explain your answer.

## Think about ...

Think about numbers less than 10.

What about numbers greater than 10?

## What if?

Is a number with 1 ten always smaller than a number with 2 tens?

When you've finished, turn to page 80.

# 1 more and 1 less

## Challenge

1 more is the same as adding 1, and 1 less is the same as subtracting 1.

Do you agree with Jake?

Explain your answer.

## Think about ...

Think about counting on and back on a number line.

Is it the same for numbers larger than 10 as it is for numbers smaller than 10?

## What if?

Is 2 more the same as adding 2, and 2 less the same as subtracting 2?

What about 10 more or 10 less? Is this the same as adding or subtracting 10?

Explain your answers.

When you've finished, turn to page 80.

## Challenge

What is the same about these three numbers?

What is different?

## Think about …

Think about the position of the 4 in each of the numbers.

What do you notice about these numbers if you count on in twos from 0? What does this mean?

## What if?

What is the same and what is different about the numbers 12 and 21?

What about the numbers 4 and 14?

What about the numbers 5 and 16?

When you've finished, turn to page 80.

# Wrong patterns

## Challenge

What is the mistake in each pattern?

| 15 | 16 | 18 | 19 | 20 |

| 13 | 12 | 11 | 9 | 8 |

| 60 | 50 | 35 | 30 | 20 |

| 6 | 8 | 9 | 12 | 14 |

What should the correct patterns be?

## Think about ...

What is the rule for each pattern?

What would each pattern look like as jumps on a number line?

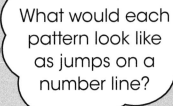

## What if?

If you count on from 16, will you say the number 12? Explain why.

If you count on in twos from 6, will you say the number 20? Explain why.

When you've finished, turn to page 80.

# Recognising halves

## Challenge

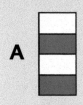
**A**

**B**

**C**

**D**

**E**

**F**

**G**

**H**

**I**

**J**

**K**

**L**

**M**

Sort these shapes into two groups.

1 Shapes that show a **half** shaded.

2 Shapes that do **not** show a half shaded.

Explain your reasoning.

## Think about …

How are you going to show how you have sorted the shapes?

Look for things that are the same and things that are different about the shapes.

## What if?

For each of your two groups, draw two other shapes that belong to the group.

When you've finished, turn to page 80.

# Finding halves

## Challenge

To find half, you put the cubes into two groups.

Ruby is explaining to Holly what it means to find half of a group of objects.

Is Ruby's explanation a good one?

How could you make it better?

## Think about ...

Give some examples in your explanation.

Does your explanation work for finding half of any group of objects?

## What if?

Holly says:

4 is half of 8.

Find other pairs of numbers where one number is a half of the other.

What do you notice about the pairs of numbers?

When you've finished, turn to page 80.

# Who's right?

## Challenge

I can use these three numbers to make four number sentences.

No you can't, you can only make one number sentence.

Who's right – Holly or Meru?

Explain your answer.

## Think about ...

Why might each child think that they are correct?

Give examples to explain your answer.

## What if?

Choose three of these number cards. Can you use them to make four number sentences?

0  1  2  3  4  5  6  7  8  9  10

What if you choose three different cards?

When you've finished, turn to page 80.

# Number sentence patterns

## Challenge

Look at this addition pattern for 8.

Work out the missing numbers.

What patterns do you notice?

Make a similar pattern for 10.

$$0 + 8 = 8$$
$$1 + 7 = 8$$
$$2 + \square = 8$$
$$3 + \square = 8$$
$$\square + 4 = 8$$
$$\square + 3 = 8$$
$$6 + \square = 8$$
$$\square + 1 = 8$$
$$\square + \square = 8$$

## Think about ...

Look at the pattern made by the first number in each number sentence. Look at the pattern made by the second number.

The pattern for 8 has nine different number sentences. How many number sentences will your pattern for 10 have?

## What if?

Make a similar pattern for 8 involving subtraction. What patterns do you notice?

Choose another number, from 9 to 12, and make another subtraction pattern.

When you've finished, turn to page 80.

# Using known facts

## Challenge

I know that 2 + 4 = 6.

How can you use 2 + 4 = 6 to work out other facts?

Write as many other facts as you can.

Explain how knowing 2 + 4 = 6 helps you know each of these other facts.

## Think about ...

Think about addition and subtraction number sentences.

What is the hardest number sentence you can make? Why is it the hardest? How does knowing 2 + 4 = 6 help you?

## What if?

Choose an addition number sentence you know. What other related facts do you know?

What if you choose a known subtraction number sentence?

When you've finished, turn to page 80.

# True statements

## Challenge

Write a number in each box to make a true statement.

$$6 + \boxed{\phantom{0}} = \boxed{\phantom{0}} + 9$$

Explain how you worked out what numbers to write.

Can you write a different number in each box and still make a true statement? Explain why.

## Think about ...

What does the = symbol mean you have to do?

What patterns can you spot to help you write different true statements?

## What if?

Write a number in each box to make this statement true.

$$\boxed{\phantom{0}} + 7 = \boxed{\phantom{0}} - 3$$

Explain how you worked out what numbers to write.

Write different numbers in each box to make other true statements.

When you've finished, turn to page 80.

# Are they right?

## Challenge

If I start at 0 and count on in twos, I'll say the number 24.

If I start at 0 and count on in tens, I'll say the number 50.

If I start at 15 and count back in twos I'll say the number 2.

Are Jake, Ruby and Holly right?

Explain your answer.

## Think about ...

Can you tell if each child is correct without having to count? If so, how?

Think about the patterns made when you count on from 0.

## What if?

Write five numbers that Jake, Ruby and Holly might each say as they count.

Write five numbers that Jake, Ruby and Holly will definitely **not** each say.

When you've finished, turn to page 80.

# Two towers

## Challenge

Take 10 cubes and use them to build two towers the same height.

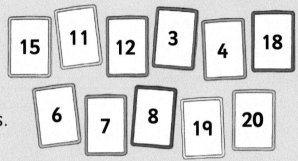

Now choose a number card and collect that number of cubes.

15  11  12  3  4  18

6  7  8  19  20

Do you think you can build two towers the same height using the number of cubes?

Repeat for the other number cards.

What do you notice?

**You will need:**
• about 30 interlocking cubes

## Think about …

What is the same about all the numbers that **can** be built into two towers the same size?

What is the same about all the numbers that **cannot** be built into two towers the same size?

## What if?

Choose numbers of your own. Can you predict which numbers you can build into two towers the same size and which numbers you can't?

When you've finished, turn to page 80.

# Biscuit tray

## Challenge

This tray of biscuits shows four facts about multiplication and division.

Do you agree with Jake?

Explain your answer.

## Think about ...

How are the biscuits grouped on the tray? Is there another way to describe this?

How does the arrangement of the biscuits show how they might be shared out?

## What if?

What are the different ways that Jake might have recorded the four facts?

When you've finished, turn to page 80.

# Doubling and halving

## Challenge

I can double any number, but I can only halve some numbers.

Do you agree with Holly's statement?

Explain your answer.

## Think about …

What do you notice about the numbers you can halve exactly?

Does your reasoning apply to any number – large and small?

## What if?

Holly also says:

Doubling is the opposite of halving.

Do you agree with Holly?

Explain your answer.

When you've finished, turn to page 80.

# Sorting 2-D shapes

## Challenge

Sort these shapes into as many different groups as you can.

Each time, explain how you have sorted the shapes.

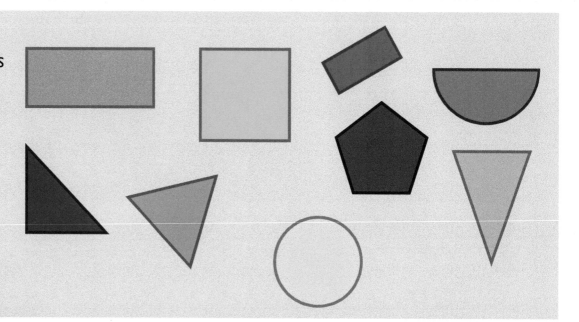

## Think about ...

What is the same about the shapes in each group?

Can you sort the shapes into two groups? What about three or four groups?

## What if?

Finding squares is easy: a square has 4 sides.

I don't think that's the best way to describe a square.

How might Jake describe a square?

When you've finished, turn to page 80.

# Same and different 3-D shapes

## Challenge

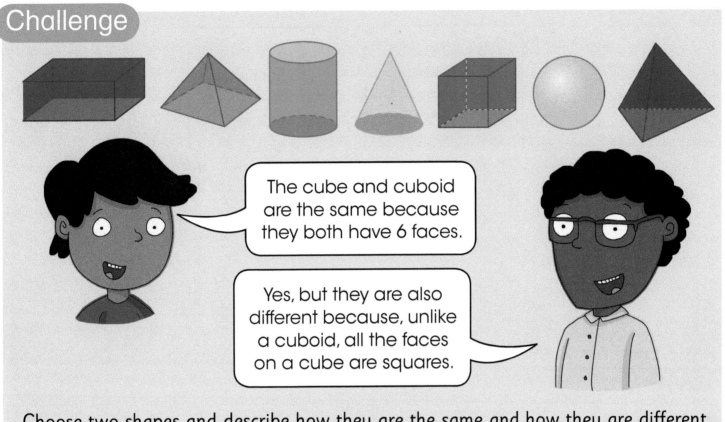

The cube and cuboid are the same because they both have 6 faces.

Yes, but they are also different because, unlike a cuboid, all the faces on a cube are squares.

Choose two shapes and describe how they are the same and how they are different.

## Think about ...

Think about the number of faces, edges and corners on the shapes.

Use words such as **square, rectangular, triangular, circular, straight** and **curved**.

## What if?

Choose three or more shapes and describe how they are the same and how they are different?

When you've finished, turn to page 80.

# Finding symmetry

## Challenge

To find a line of symmetry in a shape, I fold the shape in half.

I use a mirror to find line symmetry in a shape.

Which method do you think is best?

Explain why.

## Think about …

Think about what's good and what's not so good about each method.

Does each method work in all cases?

## What if?

How would you find line symmetry in an object in the real world such as a flower?

When you've finished, turn to page 80.

# Finding food

## Challenge

Work with a partner.

One of you secretly chooses an item in the cupboard. The other asks up to four questions to find the item. The questions can only have a 'yes' or 'no' answer.

Swap roles and repeat.

Repeat again, so that each of you has two turns of choosing and finding.

Now discuss which type of questions are good for finding the items and which type of questions aren't so good.

Explain what makes a good question.

## Think about ...

Use words such as **left**, **right**, **above**, **below**, **next to** and **first**, **second**, ...

Try to ask fewer than four questions to find the item.

## What if?

To move from the peanut butter to the flour, Jake says:

Take turns to describe to your partner how you would move from one item in the cupboard to another.

You go down two shelves and across three items to the right.

When you've finished, turn to page 80.

# Whose coins?

## Challenge

You will need:
• selection of coins (optional)

I wish I had one of each coin.

I'd like to have one 50c coin and one 25c coin.

I'd like to have all the silver coins.

I wish I could have ten 10c coins.

Whose coins would you rather have?

Explain your reasoning.

## Think about …

How much money does each child have?

Can you use counting on in steps to help you?

## What if?

Ten 5c coins is worth the same as five 10c coins.

No it's not. Ten 5c coins is worth more than five 10c coins.

Who's right? Explain why.

When you've finished, turn to page 80.

# Which is better?

## Challenge

We can use these toy cars to measure the length of this table.

I think we should use the cubes.

Whose idea do you think is better?

Explain why.

What else could you use to measure the length of the table?

## Think about …

What is the best way to measure the length of the table? Why?

What things do you need to make sure of when you're measuring the length of something?

## What if?

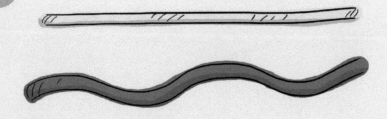

Which piece of wool is longer?
Explain your thinking.

When you've finished, turn to page 80.

# Which is heavier?

## Challenge

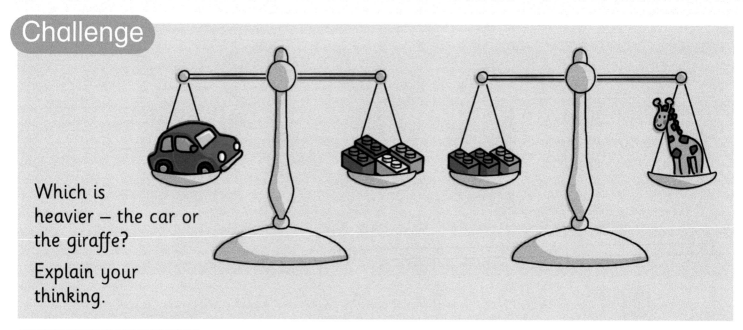

Which is heavier – the car or the giraffe?

Explain your thinking.

## Think about …

What does a balance look like when two objects weigh the same?

How does a balance show you which is the heavier object?

## What if?

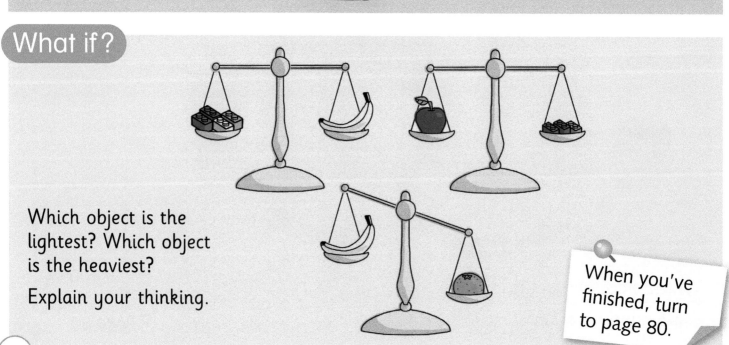

Which object is the lightest? Which object is the heaviest?

Explain your thinking.

When you've finished, turn to page 80.

# Is it fair?

## Challenge

Is Jake right?
Explain your thinking.

## Think about …

## What if?

Explain your thinking.

When you've finished, turn to page 80.

# Who's wearing which watch?

## Challenge

Our maths lesson is about to start.

It's lunchtime.

It's time for my swimming lesson.

It's my bedtime.

Who's wearing which watch?

Explain your reasoning.

## Think about ...

Each child must be wearing a different watch.

Think about the time of day that you might do each of these things.

## What if?

Could any of the children be wearing a different watch?
Explain why.

When you've finished, turn to page 80.

# Different data

## Challenge

What is the same about these two diagrams?

What's different?

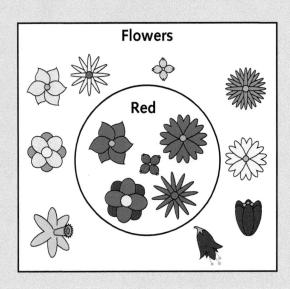

## Think about …

Think about the data in the two diagrams.

Also think about the way the data is displayed.

## What if?

What is the same about this table and the two diagrams above?

What is different?

| Red flowers | 5 |
|---|---|
| Pink flowers | 3 |
| Yellow flowers | 4 |
| Purple flowers | 2 |

When you've finished, turn to page 80.

# Finding numbers

## Challenge

  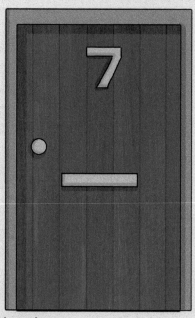

Look for numbers in your classroom and around the school.

Can you find something that shows the number 1, 2, 3, 4, …?

How many different numbers written as words can you find?

Add to your list by looking for numbers at home and elsewhere.

## Think about …

Where is a good place to look for numbers?

Think about how you're going to organise your list of numbers.

## What if?

Which number is the most common? Why?

Which is the most important number? Why?

Do you mainly see numbers written as numerals or words? Why do you think this is?

When you've finished, turn to page 80.

# List of numbers

## Challenge

1  headteacher in our school
2  eyes on my face
3  sides of a triangle
4
5
6
7
8
9
10

Meru is making a list of numbers.

Can you make your own list of numbers?

How long can you make your list?

## Think about ...

Think about things you can see and things you can't see.

You can write or draw the things on your list.

## What if?

Which numbers are easy to find? Why do you think this is?

Which numbers are hard to find? Why is this?

When you've finished, turn to page 80.

# Card patterns

Using and applying mathematics in real-world contexts

## Challenge

**You will need:**
- pack of playing cards

How many different patterns can you make using cards from a pack of playing cards?

## Think about ...

Think about:
- the colours of the cards
- the numbers on the cards
- the shapes on the cards (suits).

You don't use all 52 cards to make each pattern. You just need to make sure that you use enough cards so that each pattern repeats more than once.

Think carefully about how you're going to show your different patterns.

## What if?

Think of a pattern, then start to make the pattern using the cards. Give the remaining cards to a friend and ask them to continue the pattern.

When you've finished, turn to page 80.

# Types of shoes

## Challenge

Investigate all the different types of shoes that there are.

Now sort your list in different ways:

- what they are made from
- how you do them up
- when you wear them.

Can you sort your list in other ways?

## Think about …

Think about the special types of shoes you use for different sports or jobs.

How are you going to show the different ways you sorted your list of shoes?

## What if?

What is the most common size of shoe in your class?

Who wears the largest pair of shoes in your class?

When you've finished, turn to page 80.

# Jelly beans

## Challenge

Is the most common coloured jelly bean the most popular?

**You will need:**
- packets of jelly beans in different shapes and sizes

## Think about ...

In one packet of jelly beans, what is the most common colour? What is the least common colour?

What is the most popular colour of jelly bean? What is the least popular colour of jelly bean?

## What if?

Investigate the number of jelly beans in different packs.

Does each pack of the same size have the same number of jelly beans?

Does a pack twice as big have twice as many jelly beans in it?

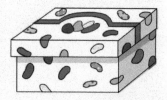
When you've finished, turn to page 80.

# Favourite bear

Using and applying mathematics in real-world contexts

## Challenge

| Floppy | Chimp | Huggy | Ted | Honey | Patch |

**You will need:**
• coloured pencils

Look at the bears on the shelf.

Which is your favourite bear? Rate each of the six bears on a scale of 1 to 10.

least favourite

favourite

1    2    3    4    5    6    7    8    9    10

Ask other children in your class to rate each of the bears.

Which bear is your class favourite?

## Think about ...

How are you going to keep of record of what each child says?

Use your class results to order the six bears, starting with the favourite.

## What if?

Investigate the different ways of dressing your favourite bear.

Your bear must wear:

• an **orange** or **purple** hat
• a **red** or **blue** T-shirt
• a **green** or **yellow** pair of shorts.

What are all the different ways of dressing your bear?

When you've finished, turn to page 80.

# Flags

Using and applying mathematics in real-world contexts

## Challenge

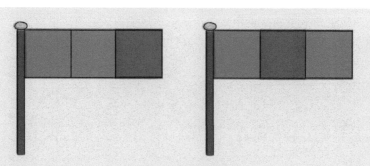

**You will need:**
- red and blue coloured pencils
- 2 cm squared paper

These two flags each have 3 red or blue squares.

Draw some more flags, each with 3 squares.

Colour the squares red or blue to make different flags.

How many different flags can you make?

## Think about ...

How many different flags have 1 red square? What about 2 red squares?

Which flags look the same turned upside down?

## What if?

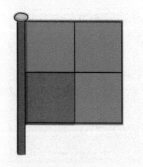 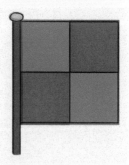

These two flags each have 4 red or blue squares.

Draw some more flags, each with 4 squares.

Colour the squares red or blue to make different flags.

How many different flags can you make?

When you've finished, turn to page 80.

62

# Letters in a name

## Challenge

I have 5 letters in my first name and 6 letters in my last name. That's 11 letters altogether.

How many letters are there in your first and last names altogether?

In your class, who has the longest first and last names together?

Who has the shortest first and last names together?

## Think about ...

Who have the longest and shortest first names in your class?

Who have the longest and shortest last names?

## What if?

Holly says:

Most first names have fewer letters in them than last names.

Is Holly right?

Holly also says:

The letter 'e' appears more times in a name than any other letter.

Is Holly right?

When you've finished, turn to page 80.

# Greg's garage

## Challenge

Greg counted 24 wheels in his garage.

All the wheels are on either cars or motorbikes.

How many different combinations of cars and motorbikes could there be?

**GREG'S GARAGE**

Open 7 days a week

## Think about ...

Greg could have only cars in his garage, or he could have only motorbikes, or he could have some of each.

Think about how you can organise the different combinations. Can you see any patterns?

## What if?

What if Greg counted 30 wheels and all the wheels were either on cars or on two- or three-wheeled motorbikes?

When you've finished, turn to page 80.

# Takeaway food

## Challenge

Which two foods would you choose for your lunch?

Each part of your food container needs to have a different type of food.

What different combinations of two foods could you have?

## Think about …

How are you going to keep a record of all the different ways?

Try to use a system to help you find all the different ways.

## What if?

What if you use this box?

Each part of the food container needs to have a different type of food.

What different combinations of four foods could you have?

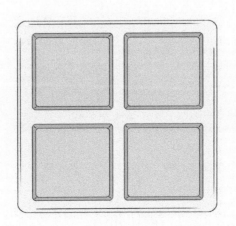

When you've finished, turn to page 80.

# Recycling

## Challenge

At home, keep a list for a week of all the things that you recycle.

After a week, look at your list and sort it in different ways.

## Think about ...

Think about what the recycled things are made from and what they are used for.

Think about how you are going to record the different ways you sorted your list.

## What if?

Jake sorted his list of recycled things by what they were made from.

He drew this diagram to show how much of each type of material was recycled.

About half of our recycling is plastic.

| Our week's recycling | | | |
|---|---|---|---|
| plastic | glass | metal | paper |

We recycle more metal than glass or paper.

Draw a diagram to show how much of each type of material you recycled.

Write statements about your recycling.

When you've finished, turn to page 80.

# Things with wheels

## Challenge

What things have wheels?

Make a list.

Sort your list into things that have 1, 2, 3, 4, ... wheels.

Now write out the list again, in order of size of their wheels, starting with the thing that has the largest wheels.

## Think about ...

Think about things that have wheels that you see at home – inside and outside.

Remember – some things have wheels that you can't see.

## What if?

Wheels are round. Make a list of other objects that are round.

When you've finished, turn to page 80.

# Pizza box

## Challenge

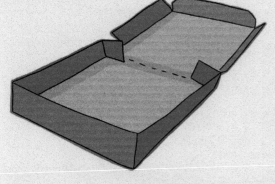
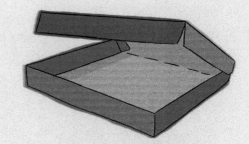
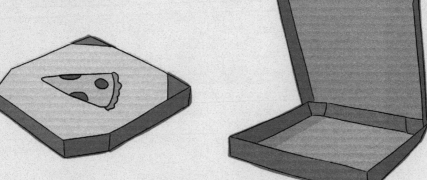

**You will need:**
- paper
- card
- ruler
- sticky tape
- glue
- scissors
- coloured pencils or crayons

Design and make a box for delivering pizzas in.

## Think about ...

Think about the shape and size of the box.

Make sure that you plan and design your box before you start to make it.

## What if?

Design a logo to put on your pizza box.

You can only use three different colours in your logo and it must be rectangular.

When you've finished, turn to page 80.

# Things that move

## Challenge

Make a list of all the things you move every day.

Organise your list under different headings.

## Think about ...

Think about things you move at home and at school.

Think about the things you:
- open
- close
- press
- slide
- push
- lift
- move down
- move out
- move up
- move in.

## What if?

What are the different shapes of all the things you move every day?

What is the most common shape you move each day?

When you've finished, turn to page 80.

# Turning letters

## Challenge

Which capital letters look the same after a half turn?

## Think about ...

Can you imagine what the letters look like rather than having to turn the letters?

Think about how you're going to show your results.

## What if?

Investigate straight and curved lines in capital letters.

Which letters have straight lines only? Which letters have curved lines only? Which letters have straight and curved lines?

Which letters fold in half exactly?

When you've finished, turn to page 80.

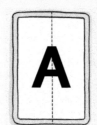

# Treasure

## Challenge

Secretly hide some treasure in the classroom.

Draw a map to show where the treasure is hidden.

Now hide your map.

Make up some clues about where the map is hidden.

**You will need:**
• an object to use as treasure

## Think about ...

You might want to use words and terms such as:

- next to
- behind
- between
- under
- below
- above

- over
- underneath
- opposite
- in front of
- to the left of
- to the right of.

Think carefully about what to include on your map.

## What if?

What if you hide the treasure and map somewhere else in the school – perhaps in another classroom or in the playground?

When you've finished, turn to page 80.

# Name's worth

## Challenge

In this challenge, each letter in the alphabet is worth an amount of money.

**A** 1c   **B** 2c   **C** 3c   **D** 4c ... **Z** 26c

So the name Jake is worth 27c.

**J** 10c   **A** 1c   **K** 11c   **E** 5c

How much is your name worth?

## Think about ...

Work out the value of each of the letters in the alphabet.

Is the name with the most letters the name that is worth the most? Is the name with the fewest letters the name that is worth the least?

## What if?

In your class, whose name do you think is worth the most?

Whose name do you think is worth the least?

Why do you think this?

What names can you think of that are worth 25c?

When you've finished, turn to page 80.

# Beads

## Challenge

**You will need:**
- collection of different types of threading beads
- wool or string
- scissors
- 1-metre length of wool or string

Make a necklace to wrap around your neck three times.

Before you make your necklace, you need to work out how long the piece of wool needs to be.

Your necklace also needs to have a repeating pattern of beads on it.

## Think about ...

You should only have beads on the front part of your necklace.

Your necklace needs to fit comfortably around your neck.

## What if?

How many beads do you need to thread onto a piece of string 1 metre long?

Estimate first, then use the beads and string to check.

When you've finished, compare your 1-metre bead string with another pair or group. Do your bead strings have the same number of beads? Why is this?

When you've finished, turn to page 80.

# How tall?

## Challenge

**You will need:**
- measuring equipment

If you measure three times around your head, the length is about how tall you are.

Is Ruby right?

## Think about ...

What, or who, do you need to help you find this out?

'About' means that your results don't have to be exactly the same – just close.

## What if?

Ruby also says:

I've just measured the length of my foot. My height is about 6 times the length of my foot.

Is Ruby right?

When you've finished, turn to page 80.

# Newspaper

Using and applying mathematics in real-world contexts

## Challenge

**You will need:**
- newspaper
- scissors
- sticky tape
- glue
- string
- measuring equipment

Using newspaper, make a model of a person that is half as tall as you are.

## Think about ...

How are you going to make sure that the model is half your size?

How are you going to keep your model together?

## What if?

Take a sheet of newspaper, open it out and place it on the floor.

Estimate, and then find out, how many children can stand on the sheet of newspaper.

How many sheets of newspaper would be needed for the whole class?

When you've finished, turn to page 80.

# Sheet of paper

Using and applying mathematics in real-world contexts

## Challenge

**You will need:**
- metre stick
- several sheets of A4 paper
- scissors
- sticky tape
- glue

Use one sheet of A4 paper to make a strip the same length as a metre stick.

## Think about ...

Think about how many small strips you will need.

Think about how wide each strip can be.

## What if?

What is the most weight a sheet of A4 paper can carry?

Think about changing the shape of the sheet of paper – perhaps by folding the paper.

How will you hold the sheet of paper so that the paper, and not your hand, is carrying the weight?

When you've finished, turn to page 80.

# Yoghurt pots

## Challenge

**You will need:**
• collection of different shaped and sized yoghurt pots

Investigate what is the same and what is different about all of the yoghurt pots.

## Think about ...

Think about:
- shape
- size
- how much each pot holds
- flavours
- price (if the pots show this)
- design and labelling.

Think about how you're going to show what you find out.

## What if?

Find two or more pots that hold the same amount of yoghurt but are different shapes.

Find two or more pots that look the same but hold different amounts of yoghurt.

Order the pots, starting with the one that holds the most yoghurt.

Which of the yoghurts would you buy? Why?

When you've finished, turn to page 80.

# Jobs rota

## Challenge

Think about all the different jobs that need doing in your class during a week.

Make a rota for all the different jobs.

Make sure that everyone has a turn at doing all the jobs.

## Think about …

Rather than making sure that each child has a turn at doing the jobs, might it be better to make sure that each group has a turn at doing the jobs?

Think about how you're going to organise and present your rota.

## What if?

Compare your rota with another pair's or group's rota.

What is the same about them? What is different?

Which is the better rota? Why?

Can you work together to create an even better rota?

When you've finished, turn to page 80.

# Cover it

## Challenge

Estimate, and then find out, how many playing cards you need to cover different things in the classroom.

**You will need:**
- pack of playing cards
- balloons
- measuring equipment

## Think about ...

Make sure that you spread the cards carefully.

Don't leave any gaps between cards and don't overlap any of the cards.

## What if?

Estimate, and then find out, how many balloons you need to cover the carpet area of the classroom.

How much space does one balloon take up?

How many balloons would you need to cover the entire floor of the classroom?

When you've finished, turn to page 80.

# When you've finished, find a partner or another pair or group.

## Share your results.

Discuss any results that are different.

Which result is correct?

Might there be more than one solution?

## Share the methods used.

Discuss the similarities and differences in the methods used.

Which method worked best?

Are there any other ways to go about solving the problem?

## Share what you have learned.

Discuss what you would do the same and what you would do differently next time.

Is there anything you would do differently?

What have you learned for next time?

What would you do the same?